YOUR KNOWLEDGE HAS VALUE

- We will publish your bachelor's and
  master's thesis, essays and papers

- Your own eBook and book -
  sold worldwide in all relevant shops

- Earn money with each sale

Upload your text at www.GRIN.com
and publish for free

Bibliographic information published by the German National Library:

The German National Library lists this publication in the National Bibliography; detailed bibliographic data are available on the Internet at http://dnb.dnb.de .

This book is copyright material and must not be copied, reproduced, transferred, distributed, leased, licensed or publicly performed or used in any way except as specifically permitted in writing by the publishers, as allowed under the terms and conditions under which it was purchased or as strictly permitted by applicable copyright law. Any unauthorized distribution or use of this text may be a direct infringement of the author s and publisher s rights and those responsible may be liable in law accordingly.

Imprint:

Copyright © 2017 GRIN Verlag
Print and binding: Books on Demand GmbH, Norderstedt Germany
ISBN: 9783668616325

This book at GRIN:

https://www.grin.com/document/378387

Prem Jose Vazhacharickal, Elizabeth Cherian, Noby Mathew

Optimization of Media Formulations for Callus Induction and direct Organogenesis in Justicia Gendrussa Burm Fil

The Uses and Advantages of an Exotic Medicinal Plant in Kerala

GRIN Verlag

GRIN - Your knowledge has value

Since its foundation in 1998, GRIN has specialized in publishing academic texts by students, college teachers and other academics as e-book and printed book. The website www.grin.com is an ideal platform for presenting term papers, final papers, scientific essays, dissertations and specialist books.

Visit us on the internet:

http://www.grin.com/

http://www.facebook.com/grincom

http://www.twitter.com/grin_com

Optimization of media formulations for callus induction and direct organogenesis in *Justicia gendrussa* Burm Fil., an exotic medicinal plant in Kerala

Prem Jose Vazhacharickal, Noby Mathew and Elizabeth Cherian

Table of contents

Table of figures

Table of tables

List of abbreviations

%	: Percentage
°C	: Degree celsius
µl	: Microliter
mL	: Millilitre
mm	: Millimetre
sp	: Species
g	: Gram
UV	: Ultra Violet
L	: Litre
2, 4-D	: 2, 4-Dichlorophenoxy Acetic Acid
BAP	: Benzylaminopurine
NAA	: α- Naphthalene acetic acid
IAA	: Indole 3-acetic acid
MS	: Murashige and Skoog
Fe EDTA	: Ferrous Ethylene diamine tetra acetic acid
mg	: Milligram
PTC	: Plant tissue culture
µg	: Microgram
nm	: Nanometre
cm	: Centimeter
FeSO4	: Iron sulfate
HCl	: Hydrochloric acid
KI	: Potassium iodide
Kn	: Kinetin
K. pneumoniae	: *Klebsiella pneumoniae*
MHA	: Muller hinton agar
mg/L	: Milligram per liter
NaCl	: Sodium chloride
NaOH	: Sodium hydroxide
P. aeruginosa	: *Pseudomonas aeruginosa*
S. aureus	: *Staphylococcus aureus*

Na$_2$ EDTA   : Disodium Ethylene diamine tetra acetic acid

B. subtilis   : *Bacillus subtilis*

Na$_2$ EDTA : Disodium Ethylene diamine tetra acetic acid

B. subtilis : *Bacillus subtilis*

Optimization of media formulations for callus induction and direct organogenesis in *Justicia gendarussa* Burm Fil., an exotic medicinal plant in Kerala

Prem Jose Vazhacharickal[1]*, Noby Mathew[2] and Elizabeth Cherian[2]
* premjosev@gmail.com
[1]Department of Biotechnology, Mar Augusthinose College, Ramapuram, Kerala, India-686576
[2]Department of Biotechnology and Applied Microbiology, St. Thomas College, Pala, Kerala, India-686574

Abstract

In this study, the callus initiation and rhizogenesis of *Justicia gendarussa* from nodal and leaf explants were established in Murashige and Skoog medium supplemented with different hormonal concentrations and also optimization of callus initiation at hormonal concentrations. As a result the callus was initiated on MS medium supplemented with different hormones. The higher concentration of NAA and BA induced callus from nodal and leaf explants. The higher concentrations of 2, 4 – D and kinetin induced callus on nodal explants. Lower concentration of 2, 4 – D and kinetin induce callus on leaf explants. The combination of NAA and kinetin induce rhizogenesis from nodal explants. The antimicrobial activity of callus was evaluated by well diffusion method. *Escherichia coli, Pseudomonas aeruginosa, Bacillus subtilus, Staphylococcus aureus, Klebsiella pneumonia* were used as test organisms. The results showed that *Justicia gendarussa* have antimicrobial activity on *Escherichia coli, Klebsiella pneumonia*, and *Staphylococcus aureus*.

Keywords: *Justicia gendarussa*, callus initiation, direct organogenesis, antimicrobial assay.

1. Introduction

Biotechnology is the advanced technology that is the use of biological processes, organisms or systems to manufacture products intended to improve the quality of human life. It has many applications in different fields such as medical, agriculture, industry, plant and animal science too. Plant tissue culture is one of the main branches of biotechnology (Alfermann and Petersen, 1995).

Plant tissue culture is the technique of maintaining and growing plant cells, tissues or organs especially on artificial medium in suitable containers under controlled environmental conditions. The part which is cultured is called as explants i.e., any part of the plant taken as explants and grown in a test tube, under sterile conditions in special nutrient media. This capacity of cell to explants to regenerate to a whole plant is called cellular totipotency. Gottileb Haberlandt was first initiated the tissue culture technique in 1902 (Alfermann and Petersen, 1995; Razdan, 2003; Pierik, 1997).

Plant tissue cultures are associated with a wide range of applications. The most important being the production of pharmaceutical, medicinal and other industrially important compounds (Razdan, 2003; Pierik, 1997).

Justicia gendarussa belongs to the family Acanthaceae. It is important medicinal plant used in the treatment of various health problems. It is commonly known as willow-leaved justicia. *Justicia gendarussa* is commonly known as Vathamkolly in Kerala. It is a small erect branched shrub. It has been described as rare and endemic to India. The plant is widely used in various forms for many of its medicinal and insecticidal properties and that it is a quick growing, evergreen forest shrub considered to be a native of China and distributed in Sri Lanka, India and Malaysia (Rahmatullah et al., 2010; Silja et al., 2008).

It is a shade loving, quick growing, erect, branched and evergreen shrub by measuring 0.6-1.2 m in height. The leaves are simple, opposite, lanceolate or linear-laneolate, acute at base, tapering into rounded apex and glabrous and shining leaves 8-12.5 cm long, 1.2-2 cm broad, with prominent purple veins beneath. The stem is quadrangular, thickened at and above the nodded and internodes measure 2-7 cm long (Rahmatullah et al., 2010; Silja et al., 2008).

The flowers are in terminal or axillary spikes and are irregular, bisexual, sessile, and white with pink or purple spots inside and red in the throat and lip (Rahmatullah et al., 2010; Silja et al., 2008).

An extract of the leaves or young shoots is used as an emetic in coughs and asthma in the Philippines, whereas fresh leaves are applied as topical to cure oedema of beri-beri and rheumatism. A decoction of the leaves is used for bathing during childbirth. In Malaysia, the leaves are much applied to treat headache and pains as a lotion to treat swellings and rheumatism, and in a bath after confinement, the roots for treating thrush and cough. The leaves are also in preparations to treat gonorrhoea, and malaria. In Indonesia, the leaves are used to treat headache, rheumatism and pain. In Vietnam, the leaves are applied externally as a poultice, decoction or tincture, to treat rheumatic arthritis and swellings. In Thailand, the roots are used against dieresis, diarrhoea. The leaves are taken internally against cough, fever and as a cardiotonic and used externally to treat inflammation, wounds and allergy (Rahmatullah et al., 2010; Silja et al., 2008; Zheng and Xing, 2009; Herrera-Mata and Rosas-Romero, 2002).

The present study was focused on the direct organogenesis, and the initiation of callus cultures from nodal and leaf explants of *Justicia gendarussa*, and optimization of the culture medium at the level of hormones. A part of the study was used to evaluate the bio control activity of the callus.

1.1 Aim

The aim of this study is the development of simple, reproducible and efficient in vitro protocol for callus proliferation and direct organogenesis of *Justicia gendarussa* in Kerala.

1.2 Objectives

The objectives of this research work were

- To initiate callus culture in Murashige and Skoog (MS) medium supplemented with different hormones
- To optimise the callus culture at the level of hormones
- To induce direct organogenesis in MS medium supplemented with different hormones
- To evaluate the antimicrobial activity of callus

2. Review of literature

Indian medicinal plants are the essence of Ayurveda and Ayurvedic treatments. When used judicially and clocking with the basic principles they produce miraculous effects. Ayurvedic drugs are rightly called the elixirs of life. Ayurvedic herbs played important role in Ayurvedic treatment, from ancient time to this most modern time. Justicia is a genus of flowering plants in the family Acanthaceae. There are 658 species found in this family (Rahmatullah et al., 2010; Silja et al., 2008; Zheng and Xing, 2009; Herrera-Mata and Rosas-Romero, 2002).

Justicia gendarussa or locally known as gendarussa have been reported to contain naringenin and kaempferol that possess antioxidant and cytotoxicity activity. Plant regeneration studies of *Justicia gendarussa* using different type of explants (young leaves and nodal segments) and media (B5 and MS basal media) supplemented with different concentration of plant growth regulators. The results indicated a maximum of callus was obtained when young leaf explants were cultured on MS medium supplemented with Kn (3.0 mg/L) + 2, 4-D (1.0 mg/L). For shoot induction, high percentage of shoot regeneration and number of shoots produced per explants was recorded when nodal explants were cultured on MSB5 plates supplemented with 2 mg/L BAP. Then the shooted explants were rooted in medium- free hormone (Ayob et al., 2012).

Antioxidant and phenolic compound was found in *Justicia gendarussa* via total phenolic content (TPC) and α, α – diphenyl-β- pycrilhydrazil hydrate (DDPH) radical scavenging assays. The assays were applied on aqueous and methanolic extracts of leaves, callus culture and cell suspension culture of *Justicia gendarussa*. Callus was induced from green and matured leaves of *Justicia gendarussa* cultured on Murashige and Skoog(MS) basal medium supplemented with different concentration of 2,4-dichlorophcnoxyacetic acid (2,4 D) or 1 naphthalenaacetic acid(NAA) showod highor phenolic content and antioxidant activity as compared to the one induced on 2,4-D. the highest total phenolic content was observed in callus culture induced with NAA and BAP, while the highest antioxidant activity was observed in the cell suspension (Azura et al., 2011).

Preliminary phytochemical analysis for the production of bioactive compounds in the in vitro grown cultures of stem and leaf derived calli of *Justicia gendarussa* Burm Fil. was carried out. The phytochemicals produced in vitro was compared with that of the stem and leaf samples. The study showed the presence of phenolics, flavanoids, alkaloids and terpenes in all the test samples. (Bhagya and Chandrashekar, 2012).

In vitro propagation of *Justicia gendarussa* in MS medium supplemented with NAA induced prolific callus on both leaf and nodal explants. Organogenic and chlorophyllous calli were produced at lower concentrations of NAA (1.0 mg/L) and BAP

(0.1 mg/L). Thick and long roots with numerous root hairs were produced with NAA (1.0 mg/L) and BAP (0.1 mg/L). Long shoots were also formed (Agastian et al., 2006). In vitro regeneration of shoots buds was obtained from culture of nodal cuttings as well as shoot regeneration from callus. The nodal cuttings differed in shoot proliferation in terms of percentage of explants that responded and average shoot length with various concentrations (4.4, 8.9, 13.3, 17.7, 22.2 µM) of 6- benzyl adenine (BA), kinetin (Kn) and thidizuron. Optimum 87% of cultures respond with an average shoot length of 4.4 cm on Murashig and Skoog (MS) medium supplemented with 17.7µM BA. Callus was induced from the mature leaf segments on MS medium supplemented with Kn (4.7, 13.9, 23.2µM) alone or in combination with 2, 4- dichlorophenoxyacetic acid (2, 4-D: 2.3µM, 4.5µM). Optimum callus induction (78%) was obtained on MS medium supplemented with 14µM Kn and 4.5µM 2, 4-D. when the callus was sub-cultured on MS medium fortified with BA (8.9, 17.7, 26.6µM) or Kn (9.3, 18.6, 27.9µM) alone or in combination with α naphthalene acetic acid (NAA: 2.7, 5.4µM), shoot regeneration was obtained. The highest response (92%) was observed on MS medium containing 17.7µM BA and 5.4µM NAA. On this medium, an average number of 12.2 shoots were obtained per responding callus. The shoots obtained from callus and nodal cuttings were rooted with a frequency of 73% on MS medium augmented with 9.8µM indole-3-acetic acid (Dennis and Hoshino, 2010).

Micropropagation of *Justicia gendarussa* from nodal segments was developed. High percentage of proliferating micro shoot cultures were obtained by placing nodal segments on Murashige and skoog medium supplemented with 4.44µM BAP, 8.88µM, 2.22µM, 2.69µM NAA respectively (85%). Individual shoots were excised and transferred onto half strength MS medium for rooting. The half MS medium augmented with 4.92µM IBA produced maximum number and maximum percentage of rooting (Johnson et al., 1970).

The regeneration of a large number of plantlets via indirect shoot organogenesis and somatic embryogenesis has been developed from the stem and leaf explants of *Justicia gendarussa* Burm Fil. The callus was efficiently induced from the explants using Murashige and Skoog (MS) medium supplemented with α- naphthalene acetic acid (NAA)+ Benzyl amino purine ((BAP) (1.0+0.1mg/L). The highest number of plantlets through indirect shoot organogenesis was obtained when the callus was sub cultured to MS medium with BAP+NAA (0.1+1.0mg/L). The maximum number of plantlets via somatic embryos was obtained in the medium with BAP+NAA

(1.0+0.1mg/L) for stem derived calli and kinetin (Kn) + NAA 2.0+0.2mg/L) for leaf derived calli. The in vitro developed shoots were rooted well in half strength MS medium supplemented with 0.5mg/L of Indole acetic acid (IAA) (Bhagya and Chandrashekar, 2013).

The pharmacological study includes preliminary phytochemical screening of alkaloids, phenols, flavonoids and saponins which were confirmed in both in vivo grown garden plants as well as invitro generated plantlets of *Justicia adhatoda*. Explants inoculated on MS medium supplemented with different combination of auxins and cytokinins were successfully employed; combination of NAA (0.5 µg/mL) and BA (2.0µg/mL) for nodal explants was found to be effective concentration to obtain plant regenerants from nodal explants. Rooting in the regenerants was also successfully initiated on transfer into half MS basal medium, supplemented with IBA (1.0µg/mL) (Bhawana et al., 2017).

The in vitro grown stem and leaf derived calli of Justicia gendarussa Burm Fil. obtained on solid MS medium supplemented with NAA+BAP(1+0.1mg/L) were cultured on solid and liquid MS medium for 32 to 35 days (Bhagya and Chandrashekar, 2013).

The effectiveness BAP and NAA growth hormones on establishment of plant regeneration for selected ornamentals: *Agapanthus praecox, Justicia betonica* and *Celosia cristata*. Various explants (leaf, stem, shoot tip and bulb) derived from one month old aseptic seedlings of *Agapanthus praecox, Celosia cristata*, as well as explants from intact plants of *Justicia betonica* were utilized to achieve complete plant regeneration of plant species. MS medium supplemented with various hormones, with an emphasis on BAP and NAA were tested to obtain direct and indirect regeneration. Both *Agapanthus praecox* (bulbs) and *Celosia cristata* (shoots) formed complete plantlets on MS added with 0.5 – 2.0mg/L BAP and NAA, while direct regeneration was achieved for *Justicia betonica* on MS media containing BAP (Jamilah et al., 2014).

Callus induction and in vitro propagation of *Justicia betonica* from three weeks old petiole and internode explants. Callus was readily induced from petiole explants when cultured on Murashige and Skoog medium (MS) fortified with NAA (0.5-2.0mg/L), BAP (0.5-2.0mg/L), kinetin (1.0-2.0mg/L) and zeatin (1.02-2.0mg/L), while internode explants only showed formation of callus on MS basal and when 0.5mg/L BAP or 1.5mg/L kinetin were added. Optimum in vitro regeneration Justicia betonica had been successfully achieved using internode explants cultured on MS medium supplemented with 1.5 mg/L NAA and 0.5mg/L BAP. Rhizogenesis was observed from petiole cultures supplemented with 1.0mg/L NAA and 1.0 mg/L BAP as well as MS with 1.5mg/L NAA and 0.5 mg/L BAP (Yaacob et al., 2013).

Studied the effect of the plant extract *Justicia spicigera* in different haematopoietic cells: human leukaemic cell lines, umbilical cord blood cells, and mouse bone marrow cells. By examining colony formation and performing the MTT (3-(4, 5- dimethylthiazol-2-)- 2, 5- diphenyl tetrazolium bromide) assay it was shown that the plant extract of *Justicia spicigera* contains cytotoxic factors for leukaemic cells and has no proliferative activity on normal haematopoietic progenitor cells. The results show that this plant extract induces apoptosis in the human leukaemia cell line TF-1 but not in the bcl-2 transfectant cell line TB-1. Similar results were obtained using a haemopoietic cell line 32D and 32DBcl2. The cultures of umblical cord blood cells and mouse bone marrow that contain granulocyte macrophage colony stimulating factor (GM-CSF) do not proliferate or become terminally differentiated in the presence of the infusion of *Justicia spicigera*. GM-CSF that acts by abrogating programmed cell death is not sufficient to inhibit the apoptotic stimulus in TF-1and 32-D cells. Moreover, mouse fibroblasts and two cervical carcinoma cell lines CALO and INBL, undergo apoptosis in the presence of different concentrations of an infusion from the plant. The data show that there is a strong correlation between the cytotoxic effect and cell proliferation. Together, these results indicate that the plant infusion of *Justicia spicigera* does not contain any haematopoietic activity, induces apoptosis inhibited by bcl-2 and is linked to cell proliferation (Caceres-cortes et al., 2001).

In vitro antioxidant activity of stem extracts of *Justicia gendaruusa* Burm along with in vivo heapatopoietic activity of methanolic fraction was carried out to ascertain the folkloric claim of its hepatoprotective activity. The crude methanolic extract was prepared by soxhlet extraction and fractional into pet ether, chloroform, and methanolic fractions. The marc remained was further extracted with double distilled water by refluxation in water bath. Preliminary phytochemical testes, total phenolic and flavonoid content present in each fraction / extract was determined. All fractions and aqueous extract were evaluated for their antioxidant activity using DPPH free radical scavenging activity, hydrogen peroxide scavenging activity, reduction of ferric ion in presence and absence of EDTA. The methanolic fraction was further studied for its in vivo hepatoprotective activity using CCl_4 induced hepatotoxicity in albino rats. The various biochemical parameters were evaluated to assess its hepatoprotective activity. Methanolic fraction has more phenolic / flavonoid content and shows good antioxidant activity. The methanolic fraction has a good hepatoprotective activity at dose of 300mg/kg-1 in albino rats. Interestingly its hepatoprotective activity, decreases as the dose increases. Stem extract of *Justicia gendarussa* has moderate hepatoprotective activity; it may be due to its total phenolic and flavonoid contents (Krishna et al., 2010).

Justicia gendarussa methanolic leaf extracts from five different locations in the southern region of Peninsular Malaysia and two flavonoids, kaempferol and naringenin, were tested for cytotoxic activity. Kaempferol and naringenin were two flavonoids detected in leaf extracts using gas chromatography flame ionization detection (GC-FID). The results indicated that highest concentrations of kaempferol and naringenin were detected in leaves extracted from mersing with 1591.80mg/kg and 444.35 mg/kg, respectively. Positive correlations were observed between kaempferol and naringenin concentrations in all leaf extracts analyzed with the Pearson method. The effects of kaempferol and naringenin from leaf extracts were examined on breast cancer cell lines (MDA-MB-231 and MDA-MB-468) using MTT assy. Leaf extracts from mersing showed high cytotoxicity against MDA-MB-468 and MDA-MB-231 with IC50 values of 23 µg/mL and 40µg/mL, respectively, compared to other leaf extracts. Kaempferol possessed high cytotoxicity against MDA-MB-468 and MDA-MB-231with IC50 values of 23 µg/mL and 34µg/Ml, respectively. These findings suggest that the presence of kaempferol in mersing leaf extract contributed to high cytotoxicity of both MDA-MB-231 and MDA-MB-468 cancer cells (zahidah et al., 2014). Analysis of methanolic infusions of leaves showed that *Justicia pectoralis* contained 1,2 –benzopyrene and umbelliferone as the main constituents, and quercetin and kaempferol as the main aglycones, while *Justicia gendarussa* contained no cumarins, and the flavonoids were c- glycosides, with α- amyrin as the main titerpenoid, followed by similarenol, β- amyrin and β- sitosterol (Oliveira et al., 2000).

To evaluate the phytochemical and antimicrobial properties of ethanolic and aqueous extracts of stem and leaves (*Justicia gendarussa*) against 12 human pathogens. Antimicrobial activity was evaluated by the disc diffusion and broth dilution methods and standard procedure were followed for the identification of phytoconstituents. The aqueous extracts of stem (*Justicia gendarussa*) showed maximum inhibitory activity against *Shigella flexineri* (26.20 mm), *Proteus mirabilis* (24.50mm), *Escherichia coli* (21.40 mm) and *Bacillus subtils* (20.25 mm) respectively and ethanolic extract showed less inhibitory activity. The aqueous extract leaves showed significant antimicrobial activity against only *Staphylococcus aureus* (26.33 mm) while the ethanolic extract of leaves showed slight inhibitory activity against a few organisms. The stem extracts (aqueous and ethanolic) showed significant antimicrobial activity against most of the human pathogens in both methods. The result revealed that the antimicrobial properties of stem and leaves of *Justicia gendarussa* moreover associated with the presence of phenolic compounds, flavonoids, terpenoids, glycosides and tannins (Subramanian et al., 2012).

2.1 *Justicia gendarussa* Burn Fil. Common and local names

Common name: Warer willow

Hindi: Nili nargandi

Tamil: Vadaikkutti

Malayalam: Karunochchi

Kanada: Aduthodagidda

Sanskrit: Bhutakeshi

2.2 Taxonomical classification: *Justicia gendarussa* Burn Fil.

Kingdom: Plantae-- planta, plantes, plants, vegetal

Phylum: Tracheophyta

Class: Magnoliopsida

Order: Lamiales

Family: Acanthaceae

Genus: Justicia

Species: *Justicia gendarussa*

Figure 1. Description of *Justicia gendarussa* Burm Fil. a) plant with leaves, b) plant with flowers, c) and d) plant with leaves and flowers, e) with flowers.
Adapted from
http://dantri.com.vn/khoa-hoc-cong-nghe/phat-hien-dot-pha-ve-tac-dung-dieu-tri-hiv-aids-cua-cay-la-lieu-tim-thay-o-viet-nam-20170621091112417.htm
www.indianmedicinalplants.info;
http://www.chhajedgarden.com/;
http://florakarnataka.ces.iisc.ac.in

3. Hypothesis

The current research work is based on the following hypothesis

1) The callus culture growth response and organogenesis of *Justicia gendarussa* Burn Fil. vary by changing the hormones concentrations.

4. Materials and Methods

4.1 Study area

Kerala state covers an area of 38,863 km^2 with a population density of 859 per km^2 and spread across 14 districts. The climate is characterized by tropical wet and dry with average annual rainfall amounts to 2,817 ± 406 mm and mean annual temperature is 26.8°C (averages from 1871-2005; Krishnakumar et al. (2009)). Maximum rainfall occurs from June to September mainly due to South West Monsoon and temperatures are highest in May and November.

4.2 Plant material

Justicia gendarussa Burm Fil. plant materials were collected and identified from the herbal garden of Botany department in St. Thomas College Pala, Kottayam (90 42'30.1680" N and 76041'5.6904"E).

4.3 Equipment's used for experiment

Sterile forceps, sterile surgical blade, sterile test tubes, beakers, conical flasks, micropipettes, pH meter, weighing balance, autoclave and hot air oven.

4.4 Culture media

A) Murashige and Skoog medium (MS Medium- 1962)

B) Nutrient broth

C) MHA (Mullar Hinton Agar)

4.5 Compositions of MS media (Chawla H.S.)
Major salts

Ammonium nitrate (NH_4NO_3)	-	1650 mg/L
Calcium chloride ($CaCl_2.2H_2O$	-	440 mg/L
Magnesium sulphate ($MgSO_4.7H_2O$)	-	370 mg/L
Potassium dihydrogen phosphate (KH_2PO_4)	-	170 mg/L
Potassium nitrate (KNO_3)	-	1900 mg/L

Minor salts

Boric acid (H_3BO_3)	-	6.2 mg/L
Cobalt chloride ($CoCl_2.6H_2O$)	-	0.025 mg/L
Ferrous sulphate ($FeSO_4.7H_2O$)	-	27.8 mg/L
Potassium iodide (KI)	-	0.83 mg/L
Sodium molybdate ($Na_2MoO_4.2H_2O$)	-	0.25 mg/L
Zinc sulphate ($ZnSO_4.7H_2O$)	-	8.6 mg/L
EDTA disodium salt ($Na_2EDTA.2H_2O$)	-	37.3 mg/L

Vitamins and organics

Myoinosiol	-	100 mg/L
Nicotinic acid	-	0.5 mg/L
Pyridoxine HCl	-	0.5 mg/L
Thiamine HCl	-	0.1 mg/L
Glycine	-	2 mg/L
Sucrose	-	30 g
Agar	-	8 g

4.6 Preparation of MS media (stock preparation)

It is not possible to weigh and mix all the constituents just before the preparation of medium. It is time consuming and a tedious job. Again, if 100 ml or 200 ml medium is to be prepared, then it is very difficult to weigh some constituents that are used in very small quantity for 1 L medium. So, it is convenient to prepare the concentrated stock solution of macro-salts, micro-salts, vitamins, amino acids and hormones. All stock solution should be stored in refrigerator and should be check visually for contamination with microorganism or precipitation of ingredients. Stock solution of vitamins, amino acid and hormones should not be stored for indefinite period and should be kept in a deep freezer chamber. The widely used culture medium was formulated by Murashig and Skoog (1962).

Table 1. Stock solution of macro-salts.

constituents	Amount in mg to be taken for stock solution (10X)	Amount in mg/L present in original solution	Final volume of stock solution (mL)
Ammonium nitrate	1650	16500	1000
Potassium nitrate	1900	19000	1000
Calcium chloride dihydrate	440	4400	1000
Potassium dehydrate phosphate	170	1700	1000
Manganese sulphate	370	3700	1000

To make 1000 mL of this stock solution, dissolve the salts one after another in 800 mL of double distilled water and then make up the volume. The solution is filtered and can be stored in refrigerator (10 - 16°C) for a long period until the solution is totally used.

Table 2. Stock solution of micro-salts (100X).

constituents	Amount in mg to be taken for stock solution	Amount in mg/L present in original solution (mL)	Final volume of stock solution (mL)	Amount in ml to be pipette for 1 L medium
Boric acid	620	62	100	1
Manhanese sulphate	2230	22.3	100	1
Zinc sulphate	80	8.6	100	1
Sodium molybdate dihydrate	2.5	0.025	100	1
Cobalt chloride	2.5	0.025	100	1
Copper sulphate	2.5	0.025	100	1

Dissolve the constituents one by one in test tube in 80 mL double distilled water and make up the final volume to 100 mL by adding distilled water.

Table 3. Stock solution of vitamins (100X).

constituents	Amount in mg to be taken for stock solution	Amount in mg/L present in original solution (mL)	Final volume of stock solution (mL)	Amount in ml to be pipette for 1 L medium
Thiamine HCl	10	0.1	100	1
Pyridoxine HCl	50	0.5	100	1
Nicotinic acid	50	0.5	100	1

Dissolve the constituents one by one in 80 mL distilled water and make up the volume to 100 mL by distilled water.

Table 4. Stock solution of potassium iodide (100X).

constituents	Amount in mg to be taken for stock solution	Amount in mg/L present in original solution (mL)	Final volume of stock solution (mL)	Amount in ml to be pipette for 1 L medium
KI	83	0.83	100	1

Dissolve 83 g of KI in 80 mL of distilled water and make up the final volume to 100 L by double distilled water.

Table 5. Stock solution of iron (10X).

constituents	Amount in mg to be taken for stock solution	Amount in mg/L present in original solution (mL)	Final volume of stock solution (mL)	Amount in ml to be pipette for 1 L medium
Na_2EDTA	373	37.3	100	10
$FeSO_4.7H_2O$	278	27.8	100	10

Dissolve first Na_2EDTA in 80 ml distilled water and add $FeSO_4$ and keep in a magnetic stirrer for at least 1 hour in warm condition until the colour of solution changes to golden yellow. Finally make up into final volume by adding 10 mL distilled water and store in amber coloured bottle.

Table 6. Stock solution of glycine (10X).

constituents	Amount in mg to be taken for stock solution	Amount in mg/L present in original solution (mL)	Final volume of stock solution (mL)	Amount in ml to be pipette for 1L medium
Glycine	2	0.020	100	10

20 mg of Glycine is dissolved in 80 mL of distilled water and make up to 100 mL while adding.

Table 7. Stock solution of mesoinositol (100X).

constituents	Amount in mg to be taken for stock solution	Amount in mg/L present in original solution (mL)	Final volume of stock solution (mL)	Final volume of stock solution (mL)
Mesiinositol	100	1000	100	10

Dissolve 1000 mg/L mesoinositl to 80 mL distilled water and make up into 100 ml while adding.

Table 8. Stock solution of hormones.

hormone	Required amount of stock solution (mg)	Amount of solvent required to dissolve	Amount of water to be added	Final concentration
AUXIN				
2,4-D	10	1 ml ethyl alcohol	9 mL	1 mg/mL
IAA	10	,,	,,	,,
NAA	10	,,	,,	,,
IBA	10	,,	,,	,,
CYTOKININS				
Kinetin	10	1ml HCl	9 mL	,,
BAP	10	,,	,,	,,
Zeatin	10	,,	,,	,,

4.7 Compositions of nutrient broth

Beef extract	-	3 g
Peptone	-	5 g
NaCl	-	5 g
Distilled water	-	1000 mL

4.8 Preparation of nutrient broth

Dissolve the required amount of beef extract, peptone and NaCl in distilled water and sterilize the medium in autoclave.

4.9 Compositions of MHA

Beef extract	-	2 g
Acid hydrolysate of casein	-	17.5 g
Starch	-	1.5 g
Agar	-	18 g
Distilled water	-	1000 mL

4.10 Preparation of MHA

- Suspend 38g of medium in 1 L of distilled water.
- Mix thoroughly
- Heat with frequent agitation and boil for 1 minute to completely dissolve the compounds
- Autoclave at 121^0C for 15 minutes
- Cool to 45^0C
- Pour cooled Muller Hinton Agar into sterile Petri dishes.
- Allow to solidify at room temperature

4.11 Test organisms

The pathogenic bacteria species were collected from the department of applied microbiology, St. thomas College, Pala. Bacterial strains are:-

- *Staphylococcus aureus*
- *Klebsiella pneumonia*
- *Escherichia coil*
- *Pseudomonas aeruginosa*
- *Bacillus subtilis*

4.12 Experimental systems

- ❖ Preparation and sterilization of media

- ❖ Sterilization of glass wares

- ❖ Initiation of callus in MS media

- ❖ Initiation of organogenesis in MS media

- ❖ Optimization of MS media for callus initiation

4.13 Preparation of MS media

The working media were prepared by combining the required amounts of stock solutions, adding sucrose and mesoinositol as solids. Then add the plant hormones. The auxines and cytokinins used in the media were first treated into solution with ethanol and 1N HCl. The stocks were stored in a refrigerator. After the addition of hormones adjust the pH of the media to 5.8 by adding 0.1N NaOH or 0.1N HCl, and adding agar. The agar was melted by heating. Then the culture media were transferred after sterilization to a culture tubes (35 mL in one tube). The tubes were capped with non-absorbent cotton plugs.

4.14 Preparation of explants

The young leaves and nodal regions from the plants were used as explants.

4.15 Surface sterilization of explants

The explants were soaked in tap water for 1 hour. Then they washed with distilled water for 1 minute; followed by treatment with 0.1% Megastin® (carbendazim-fungicide) for 10 minutes, and again washed thrice in distilled water. Explants were further surface sterilized with 0.1% Mercury Chloride for 5 minutes. Then washed 4 to 5 times with sterile distilled water. The sterilized explants were cut into approximately 1 cm pieces by using sterile surgical blade and carefully inoculated on the medium under a sterile condition.

4.16 Inoculation procedure

There is a high risk of contamination of the nutrient medium at the time of transfer of explants into the culture medium. Therefore, inoculation has to be carried out aseptically. The aseptic inoculation of the explants into the culture medium involves following steps.

a)   Sterilization of the transfer area
b)   Transfer of the explants.

4.16.1 Sterilization of the transfer area

Laminar Air Flow (LAF) chamber was used for inoculation. LAF chamber was sterilized by mopping the floor of the chamber with spirit. The sterile Petri dishes, forceps and knife holder with sterile blade were kept inside the chamber and were flame sterilized. The whole chamber was exposed to UV light for 30 min. Sufficient time was also given to all the heat sterilized equipment's to get cooled before inoculation.

4.16.2 Surface sterilization of explants

The explants were soaked in tap water for 1 hour. Then they washed with distilled water for 1 minute; followed by treatment with 0.1% Megastin® (carbendazim-fungicide) for 10 minutes, and again washed thrice in distilled water. Explants were further surface sterilized with 0.1% Mercuric Chloride for 5 minutes. It was then washed 4 to 5 times with sterile distilled water. The sterilized explants were cut into approximately 1 cm pieces by using sterile surgical blade and carefully inoculated on the medium under a sterile condition.

4.16.3 Transfer of the explants

Maximum precaution was taken during the time of transfer of explants. The surface sterilized explants were cut into required size, & inoculated on pre-sterilized media in the presence of a spirit lamp; Non-absorbent cotton plug wrapped in gauze cloth was used to plug the culture tubes.

4.16.4 Inoculation

Under aseptic conditions, explants were inoculated on basal MS (Murashige and Skoog) (1962) medium containing 3% (w/v) sucrose, supplemented with four different auxins (2, 4-D, NAA, IAA, IBA) and cytokinine (BA). The pH was adjusted to 5.8 prior to the addition of 0.8% agar and autoclaved at 121°C for 15 minutes. Cultures were then incubated at $20 \pm 2°C$ with a 16-hour photoperiod by cool white fluorescent tubes (Das et al. (2010) and 70-75% relative humidity (Mukherjee et al. (1991).

4.16.5 Establishment of callus cultures from nodal and leaf explants

The surface sterilized explants were cut into approximately 1 cm pieces by using sterile surgical blade. The explants were carefully inoculated on the ms medium supplemented with different hormones (NAA- 0.1,0.2,0.5, and 0.7), (BA-0.01,0.05,0.1,0.5), (2,4- D- 0.1,0.3,0.5), (kinetin-0.1,0.3,0.5).

Table 9. Concentrations of combinations taken for present study.

SI.No	Medium	Hormone with concentration(mg/L)	
		NAA	BA
1	MS Medium	0.1	0.01
2	MS Medium	0.2	0.05
3	MS Medium	0.5	0.1
4	MS Medium	0.7	0.2

After the inoculation, the culture tubes were incubated at 20±2°C, 16 hour photoperiod and 70- 75% relative humidity.

Table 10. The concentration of hormones taken for present study

SI.No.	Medium	Hormone with concentration(mg/L)	
		2,4- D	kinetin
1	MS Medium	0.1	0.1
2	MS Medium	0.3	0.3
3	MS Medium	0.5	0.5

4.16.6 Establishment of direct organogenesis from nodal explant

The surface sterilized explants of were cut into approximately 1cm pieces by using sterile surgical blade. The explants were carefully inoculated on the MS medium supplemented with different hormones. The hormones used for direct organogenesis are given below.

Table 11. Combinations of NAA and kinetin used for direct organogenesis

SI.No.	Medium	Hormone with concentration(mg/L)	
		NAA	kinetin
1	MS Medium	0.2	0.1
2	MS Medium	0.5	0.3
3	MS Medium	0.7	0.5

After the inoculation, the culture tubes were incubated at 20 ± 2°C, 16 hour photoperiod and 70- 75% relative humidity.

4.16.7 Optimization of callus initiation

In order to optimize the callus initiation in hormonal concentrations, different systems were subjected to analysis. For this purpose, MS medium were prepared with different concentrations of hormones and inoculated medium were incubated at 20 ± 2°C, 16 hour photoperiod and 70- 75% relative humidity.

27

Table 12. Combinations of NAA and kinetin used for direct organogenesis

SI.No.	Medium	Hormone with concentrations(mg/L)			
		NAA	BA	2,4-D	kinetin
1	MS Medium	0.1	0.01	-	-
2	MS Medium	0.2	0.05	-	-
3	MS Medium	0.5	0.1	-	-
4	MS Medium	0.7	0.2	-	-
5	MS Medium	-	-	0.1	0.1
6	MS Medium	-	-	0.3	0.3
7	MS Medium	-	-	0.5	0.5

4.16.8 Evaluation of antimicrobial activity

Nutrient broth was prepared and sterilized. A loop full of test organism was inoculated into each tube aseptically. It was then incubated at 37°C for 5 hours.

Antimicrobial activity was determined by agar well diffusion method. Sterile Muller Hinton agar medium was poured onto sterile petriplates. After solidification, with help of sterile cotton swabs previously inoculated culture of organisms were swabbed onto the plates. Using sterile well cutter, wells were cut in each plate. 150µl of ethanol was used as control and 150µl the ethanolic extract of *Justicia gendarussa* calllus was introduced into separate wells. The plates were then incubated at 37°C for 24 hours for the occurrence of zone of inhibition. The diameter of zone of inhibition was measured in mm.

The method involved diffusion of drugs from vertical through the solidified agar layer of a petriplates to such an extent that growth of microorganisms was prevented entirely in a circular areas or zone around well containing sample.

4.17 Statistical analysis

The survey results were analysed and descriptive statistics were done using SPSS 12.0 (SPSS Inc., an IBM Company, Chicago, USA) and graphs were generated using Sigma Plot 7 (Systat Software Inc., Chicago, USA).

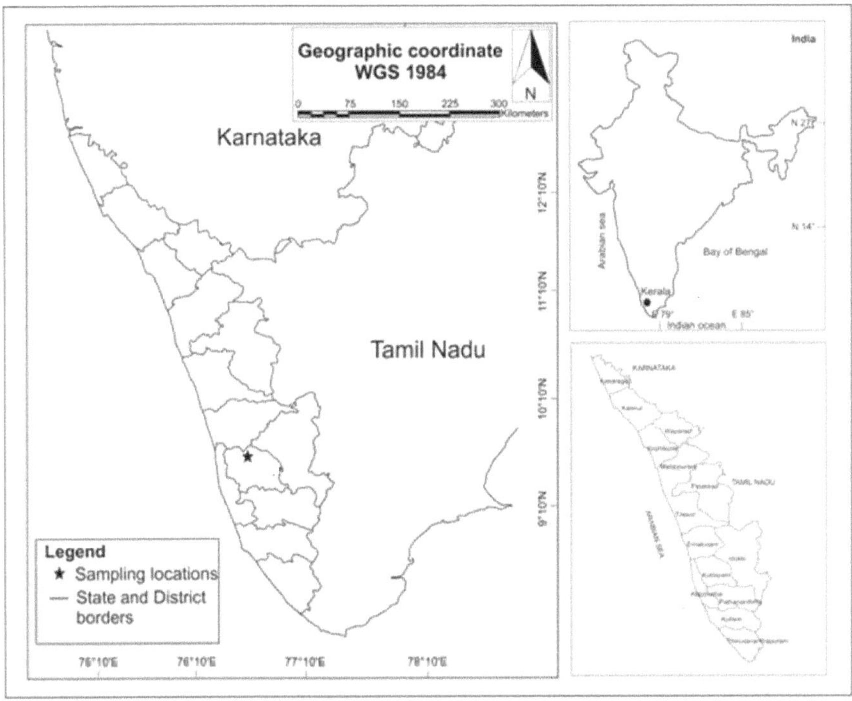

Figure 2. Map of Kerala showing the sample collection point. Authors own work.

Figure 3. Details of *Justicia gendarussa* a), explant showing the induction of callus b) and c, callus induction on MS medium supplemented with 0.7 mg/L NAA + 0.2 mg/L BA d) and e), callus induction on medium supplemented with 0.5 mg/L NAA and 0.1 mg/L BA (leaf explant). Authors own images.

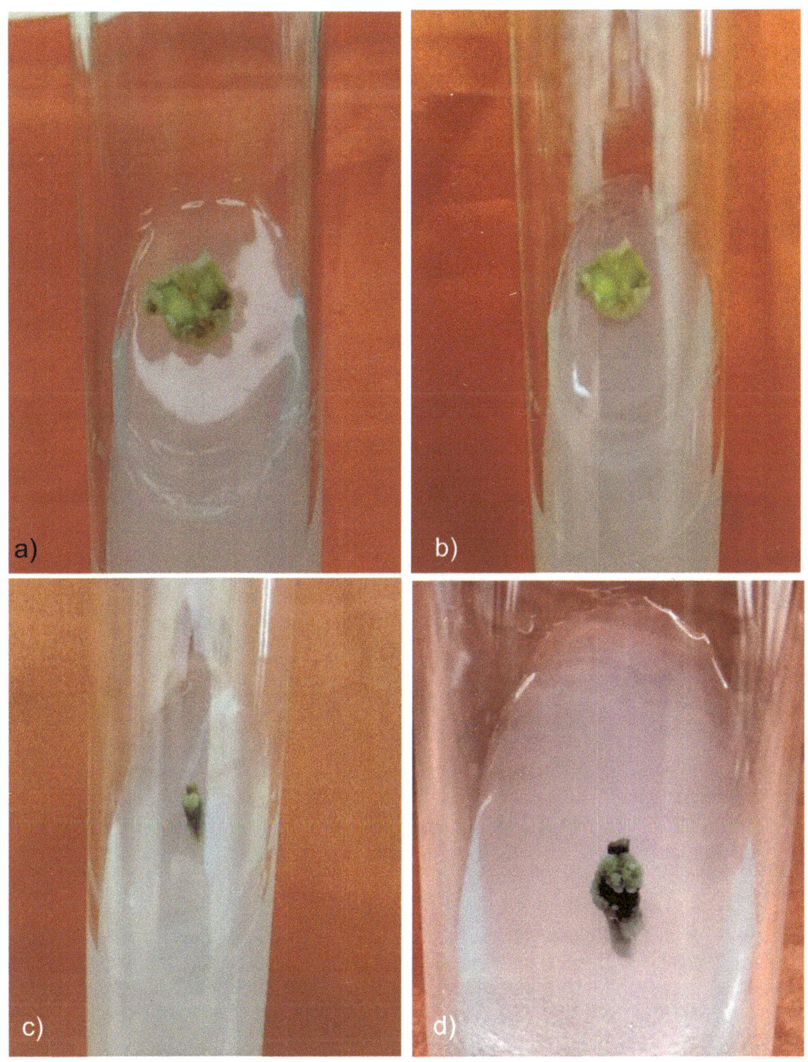

Figure 4. Details of *Justicia* gendarussa Burm Fil. a) and b) different stages of callus induction on MS medium supplemented with 0.1 mg/L 2, 4-D and 0.1 mg/L kinetin (leaf explant) c) and d) different stages of callus on MS medium supplemented with 0.5 mg/L 2, 4-D and 0.5 mg/L kinetin (nodal explant). Authors own images.

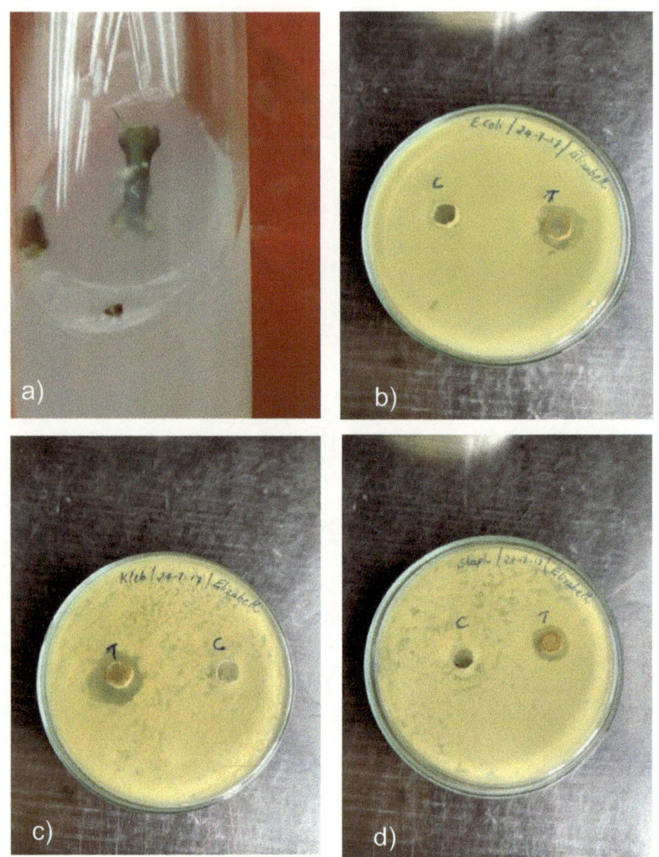

Figure 5. Details of *Justicia gendarussa* Burm Fil. a), rhizogenesis from nodal explants (0.7 mg/L NAA + 0.5 mg/L kinetin) b), MHA plate with *E. coli* (C-control; T-ethanolic extract of callus c), MHA plate with *K. pneumoniae* (C-control; T-ethanolic extract of callus d), MHA plate with *S. aureus* (C-control; T-ethanolic extract of callus. Authors own images.

5. Results and discussion

5.1 Establishment of callus cultures from nodal and leaf explants

Callus cultures were established from nodal and leaf explants of *Justicia gendarussa* on MS media supplemented with different hormones.

The production of callus and their growth behaviour was affected by the nature of explants, growth regulators and culture conditions. After 4 days of inoculation the explants showed enlargement in size. Callus initiation started from the cut ends and gradually whole callus tissue increased in size. Callus formation was observed from the 8th day of onwards.

Table 13. Growth response of leaf and nodal explants of *Justicia gendarussa* on MS medium with different hormone combinations (NAA+BA).

Sl.No	Medium	Hormone with concentration(mg/L)		Response After 2 weeks	
		NAA	BA	Nodal explants	Leaf explants
1	MS Medium	0.1	0.01	-	-
2	MS Medium	0.2	0.05	-	-
3	MS Medium	0.5	0.1	-	+ + +
4	MS Medium	0.7	0.2	+ + +	-

(+++ very good response, ++ moderate callusing, - no/very low response respectively)

From the data interpretation it was clear that, the higher concentration of NAA and BA suited for good response in both nodal and leaf explants. He lower concentration of NAA and BA inhibit the callus initiation. The calli formed from nodal explants were creamish to brown and compact. But in leaf explants, the calli were green and soft.

Table 14. Growth response of leaf and nodal explants of *Justicia gendarussa* on MS medium with different hormone combinations.

SI.No.	Medium	Hormone with concentration(mg/L)		Response after 2 weeks	
		2,4- D	kinetin	Nodal explants	Leaf explants
1	MS Medium	0.1	0.1	-	+ +
2	MS Medium	0.3	0.3	+	-
3	MS Medium	0.5	0.5	+ + +	-

(+++ Very good response, ++ Moderate callusing, + Mearge callusing, - No/very low response)

From the data interpretation it was clear that, the lower concentration of 2, 4- D and kinetin induced callus initiation in leaf explants and higher concentrations of 2, 4- D and kinetin induced callus initiation from nodal explants. The calli formed from leaf explants were cremish, while from nodal explants were green.

5.2 Establishment of direct organogenesis from nodal explants

The Direct organogenesis was established from nodal explants of *Justicia gendarussa* on MS media supplemented with different hormone combinations. The results were recorded.

The production of organs (shoot and root) and their growth was affected by the nature of explants, growth regulators and culture conditions. After 10 days of inoculation, the root was started to developed (rhizogenesis) from the nodal explants.

Table 15. Growth response (rhizogenesis) of nodal explants *Justicia gendarussa* on MS medium with different hormones.

SI.No.	Medium	Hormone with concentration(mg/L)		Response after 2 weeks (nodal explants)
		NAA	kinetin	
1	MS Medium	0.2	0.1	-
2	MS Medium	0.5	0.3	-
3	MS Medium	0.7	0.5	+ + +

- No/very low response   + Mearge rooting

+ + Moderate rooting   + + + Very good response

From the data interpretation it was clear that, the higher concentration of NAA and kinetin (0.7mg/l NAA+0.5 mg/l kinetin) induced rhizogenesis from nodal explants.

5.3 Optimization of callus initiation of *Justicia gendarussa*

For optimization of callus initiation in hormonal concentration, different systems were subjected to analysis.

Table 16. Growth response (rhizogenesis) of nodal explants *Justicia gendarussa* on MS medium with different hormones.

SI. No.	Medium	Hormone with concentrations(mg/L)				Response	
		NAA	BA	2,4-D	kinetin	Nodal explants	Leaf explants
1	MS Medium	0.1	0.01	-	-	-	-
2	MS Medium	0.2	0.05	-	-	-	-
3	MS Medium	0.5	0.1	-	-	-	+ + +
4	MS Medium	0.7	0.2	-	-	+ + +	-
5	MS Medium	0.8	0.3	-	-	+	-
6	MS Medium	-	-	0.1	0.1	-	+ +
7	MS Medium	-	-	0.3	0.3	+	-
8	MS Medium	-	-	0.5	0.5	+ + +	-
8	MS Medium	-	-	0.6	0.6	+	-

From the data interpretation it was clear that, the optimized concentration of hormones for callus induction was the combinations of 0.7mg/L NAA+ 0.2 mg/L BA and 0.5mg/L 2,4–D + 0.5 mg/L kinetin. More over the other concentrations inhibit the callus initiation.

5.4 Evaluation of antimicrobial activity

After the overnight incubation, the zone of incubation was observed in MHA plates with *Escherichia coli*, *Staphylococcus aureus* and *Klebsiella pneumoniae* against the tested solvent extracts.

Table 17. Evaluation of antimicrobial activity.

Test organism	Solvent extract	Zone of inhibition(mm)		Inference
		Control	Test sample	
E. coli	ethanol	0	20	Positive response
P. aeruginosa	ethanol	0	0	resistant
B. subtilis	ethanol	0	0	resistant
S. aureus	ethanol	0	15	Positive response
K. pneumoniae	ethanol	0	25	Positive response

From the data interpretation it was clear that, the callus extract of *Justicia gendarussa* have antimicrobial activity against *Escherichia coli, Staphylococcus aureus,* and *Klebsiella pneumonia.*

6. Conclusions

Plant tissue culture is a technique used for cell, tissue or organ culture under aseptic conditions. The technique based on the principle called totipotency that is the capacity of a cell to regenerate a whole plant. It has acquired many practical applications in agriculture and horticulture.

Justicia gendarussa is a medicinal plant and belongs to the family Acanthaceae. It is used to treat various health problems such as coughs and asthma. It is folk medicine for rheumatism.

In the present work, the young nodes and leaves of *Justicia gendarussa* were used as explants for regeneration studies and callus initiation in Murashige and Skoog medium with a pH of 5.8. Experiments were done in various concentration and combinations of auxins (NAA, 2, 4-D) and cytokinins (BA, kinetin) in order to find out an optimized medium for callus initiation. From this study, the optimized concentration of various hormones for the callus initiation was found as a combination of 0.7mg/L NAA + 0.2 mg/L BA, and 0.5 mg/L 2,4 –D + 0.5 mg/L kinetin in M S medium. The antimicrobial assay showed that the ethanolic extract of callus have some active components against the tested organisms (*Escherichia coli, Staphylococcus aureus,* and *Klebsiella pneumonia*).

Future study is required to evaluate the phytochemical analysis of the callus and develop protocol for organogenesis from other parts of the plant.

References

Agastian, P., Lincy, W., & Ignacimuthu, S. (2006). In vitro propagation of *Justicia gendarussa* Burm. f: a medicinal plant. *Indian Journal of Biotechnology*, 5(1), 246-248.

Alfermann, A. W., & Petersen, M. (1995). Natural product formation by plant cell biotechnology. *Plant Cell, Tissue and Organ Culture*, 43(2), 199-205.

Amid, A., Johan, N. N., Jamal, P., & Zain, W. N. W. M. (2011). Observation of antioxidant activity of leaves, callus and suspension culture of *Justicia gendarusa*. *African Journal of Biotechnology*, 10(81), 18653-18656.

Ayob, Z., Mohd Bohari, S. P., Abd Samad, A., & Jamil, S. (2014). Cytotoxic activities against breast cancer cells of local *Justicia gendarussa* crude extracts. *Evidence-Based Complementary and Alternative Medicine*, 2014(1), 1-12.

Ayob, Z., Saari, N. M., & Samad, A. A. (2012, July). In vitro propagation and flavonoid contents in local *Justicia gendarussa* Burm. F. In Proceedings of the 11th International Annual Symposium on Sustainability Science and Management (pp. 403-409).

Bhagya, N., & Chandrashekar, K. R. (2013). Evaluation of plant and callus extracts of *Justicia gendarussa* burm. f. for phytochemicals and antioxidant activity. *International Journal of Pharmacy and Pharmaceutical Sciences*, 5(2), 82-85.

Bhagya, N., & Chandrashekar, K. R. (2013). in vitro production of bioactive compounds from stem and leaf explants of *Justicia gendarussa* Burm. F. *Asian Journal of Pharmaceutical and Clinical Research*, 6(1), 100-103.

Bhagya, N., Chandrashekar, K. R., Karun, A., & Bhavyashree, U. (2013). Plantlet regeneration through indirect shoot organogenesis and somatic embryogenesis in *Justicia gendarussa* Burm. f., a medicinal plant. *Journal of plant biochemistry and biotechnology*, 22(4), 474-482.

Bhawna, M., Kumar, S., & Kumar, R. (2017). Pharmacognostical studies of in vivo grown garden plant and in vitro generated plantlets from nodal explants of *Justicia adhatoda* L. *International Journal of Bioassays*, 6(02), 5230-5235.

Caceres-Cortes, J. R., Alvarado-Moreno, J. A., Waga, K., Rangel-Corona, R., Monroy-Garcia, A., Rocha-Zavaleta, L., & Brousseau, R. (2001). Implication of tyrosine kinase receptor and steel factor in cell density-dependent growth in cervical cancers and leukemias. *Cancer Research*, 61(16), 6281-6289.

Cáceres-Cortés, J. R., Cantú-Garza, F. A., Mendoza-Mata, M. T., Chavez-González, M. A., Ramos-Mandujano, G., & Zambrano-Ramírez, I. R. (2001). Cytotoxic activity of *Justicia spicigera* is inhibited by bcl-2 proto-oncogene and induces apoptosis in a cell cycle dependent fashion. *Phytotherapy Research*, 15(8), 691-697.

Herrera-Mata, H., & Rosas-Romero, A. (2002). Biological activity of "sanguinaria"(*Justicia secunda*) extracts. *Pharmaceutical Biology*, 40(3), 206-212.

Johnson, M., Manickam, V. S., Das, S., & Nikhat Yasmin, A. N. (1970). In Vitro Multiplication of Two Economically Important and Endangered Medicinal Plants-*Justicia gendarussa* Drum and *Adenia hondala* (Gaertn) De Wilde. *Malaysian Journal of science*, 23(2), 49-53.

Krishna, K. L., Mruthunjaya, K., & Patel, J. A. (2010). Antioxidant and hepatoprotective potential of stem methanolic extract of *Justicia gendarussa* Burm. *International Journal of Pharmacology*, 6(2), 72-80.

Krishnakumar, K. N., Rao, G. P., & Gopakumar, C. S. (2009). Rainfall trends in twentieth century over Kerala, India. *Atmospheric Environment*, 43(11), 1940-1944.

Oliveira, A. F. M., Xavier, H. S., Silva, N. H., & Andrade, L. H. C. (2000). Chromatographic screening of medicinal Acanthaceae: *Justicia pectoralis* Jacq. and *J. gendarussa* Burm. *Revista Brasileira de Plantas Medicinais*, 3(1), 37-41.

Pierik, R. L. M. (1997). In vitro culture of higher plants. Springer Science & Business Media.

Rahmatullah, M., Kabir, A. A. B. T., Rahman, M. M., Hossan, M. S., Khatun, Z., Khatun, M. A., & Jahan, R. (2010). Ethnomedicinal practices among a minority group of Christians residing in Mirzapur village of Dinajpur District, Bangladesh. *Advances in Natural and Applied Sciences*, 4(1), 45-51.

Razdan, M. K. (2003). Introduction to plant tissue culture. Science Publishers.

Silja, V. P., Varma, K. S., & Mohanan, K. V. (2008). Ethnomedicinal plant knowledge of the Mullu kuruma tribe of Wayanad district, Kerala. *Indian Journal of Traditional Knowledge*, 7(4), 604-612.

Subramanian, N., Jothimanivannan, C., & Moorthy, K. (2012). Antimicrobial activity and preliminary phytochemical screening of *Justicia gendarussa* (Burm. f.) against human pathogens. *Asian Journal of Pharmaceutical and Clinical Research*, 5(3), 229-233.

Yaacob, J. S., Saleh, A., Elias, H., & Abdullah, S. (2014). In vitro regeneration and acclimatization protocols of selected ornamental plants (Agapanthus praecox, *Justicia betonica* and *Celosia cristata*). *Sains Malaysiana*, 43(5), 715-722.

Yaacob, J. S., Taha, R. M., Jaafar, N., Hasni, Z., Elias, H., & Mohamed, N. (2013). Callus induction, plant regeneration and somaclonal variation in in vivo and in vitro grown White shrimp plant (*Justicia betonica* Linn.). *Australian Journal of Crop Science*, 7(2), 281-288.

Zheng, X. L., & Xing, F. W. (2009). Ethnobotanical study on medicinal plants around Mt. Yinggeling, Hainan Island, China. *Journal of Ethnopharmacology*, 124(2), 197-210.

YOUR KNOWLEDGE HAS VALUE

- We will publish your bachelor's and
  master's thesis, essays and papers

- Your own eBook and book -
  sold worldwide in all relevant shops

- Earn money with each sale

Upload your text at www.GRIN.com
and publish for free